SUPER-FAST ANIMALS

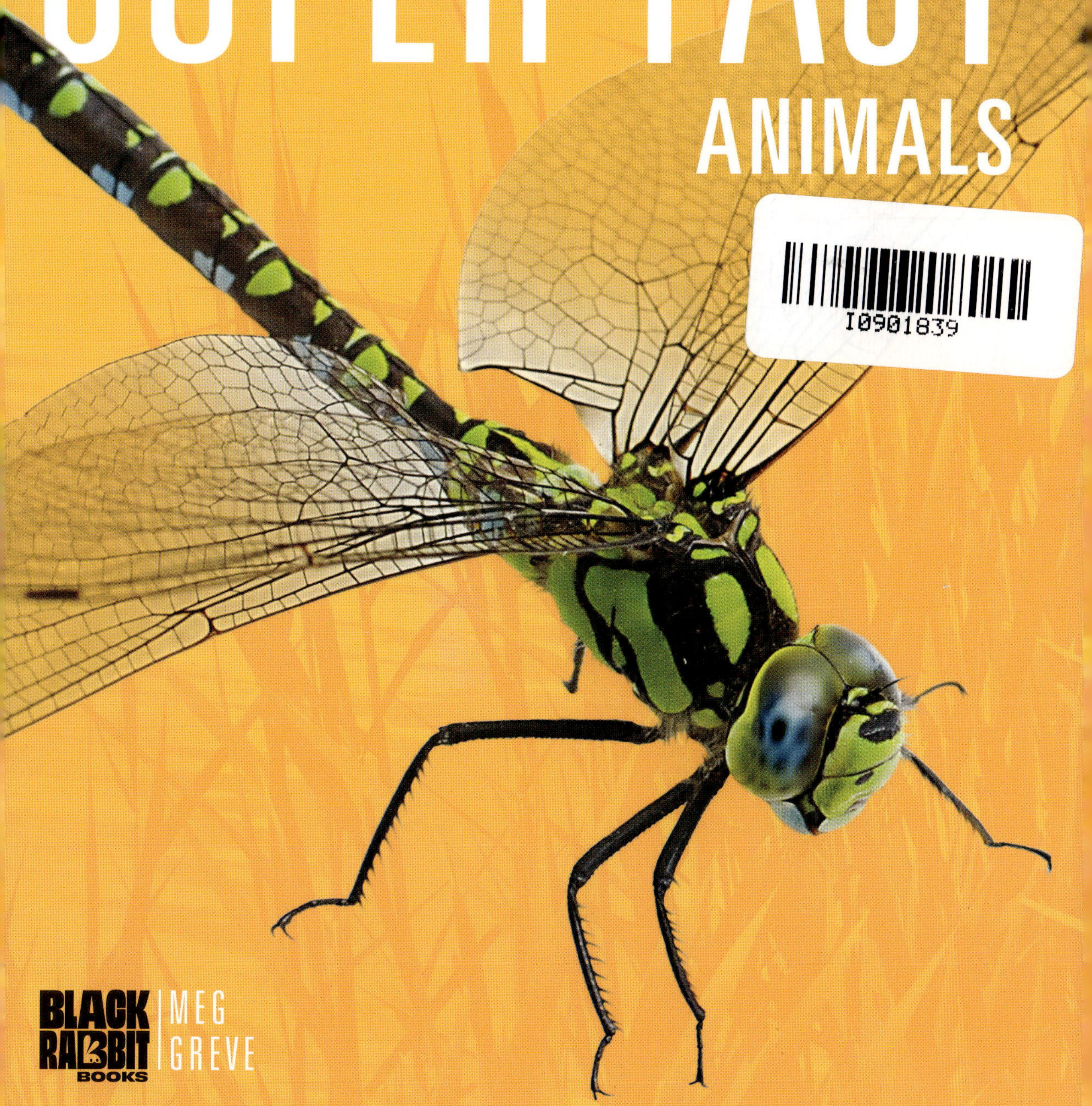

I0901839

BLACK RABBIT BOOKS

MEG GREVE

TABLE OF CONTENTS

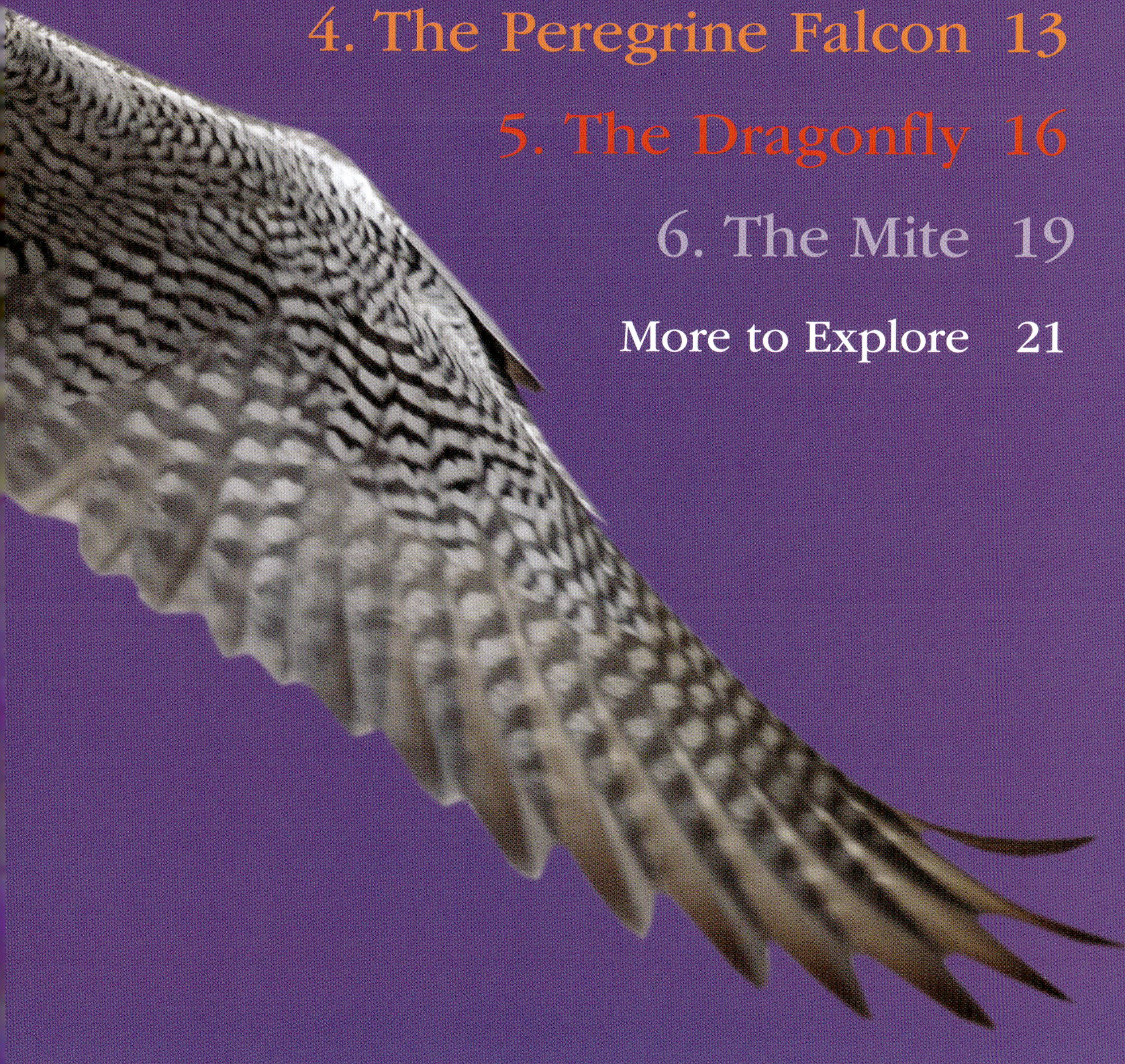

1

The Cheetah

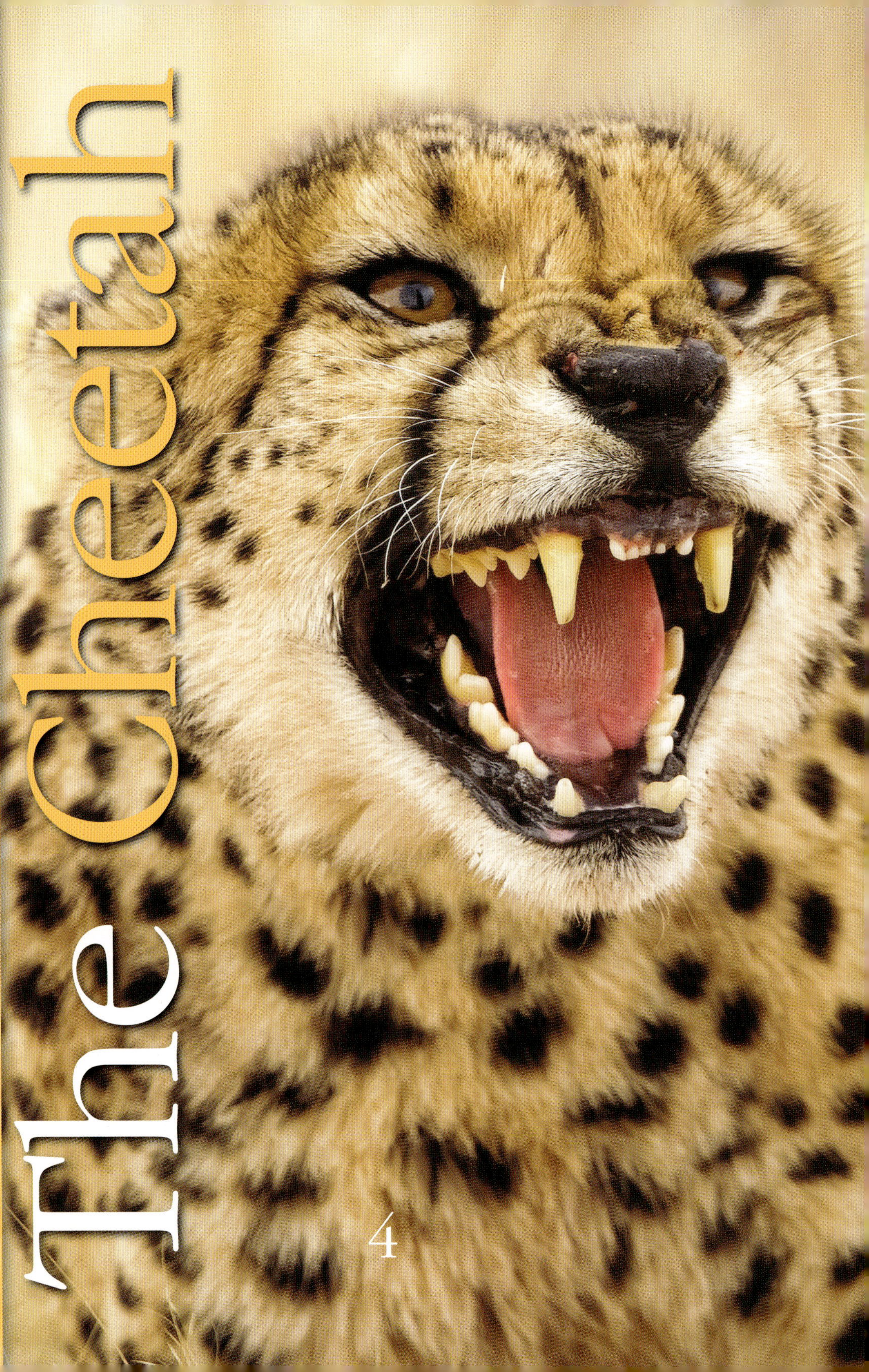

What's the fastest land animal? A cheetah! It can reach speeds up to 75 miles (121 kilometers) per hour. How does it do this?

A cheetah's spine stretches. Its legs reach to cover more ground. Its claws dig into the dirt so it does not slip. And its long tail helps it balance. The cheetah can turn without slowing down.

Cheetahs eat rabbits and antelopes. These animals are fast too. To catch them, a cheetah first sneaks up. Then it gives chase. These cats reach top speeds quickly. But they can only run fast for about a minute.

Think About It

Cheetahs are built for speed. Which traits do you think are most important to run fast?

2

The Pronghorn

Which animal is the second fastest on land? That would be the pronghorn! It can run up to 60 miles (97 km) per hour. It is the fastest land animal in North America.

Pronghorns can outrun most of their **predators**. They can spot a coyote or wolf far away. When they do, the white hair on their back rises. The rest of the herd can see the warning. Then they all run.

Did You Know?
Pronghorns migrate over 300 miles (483 km) each year!

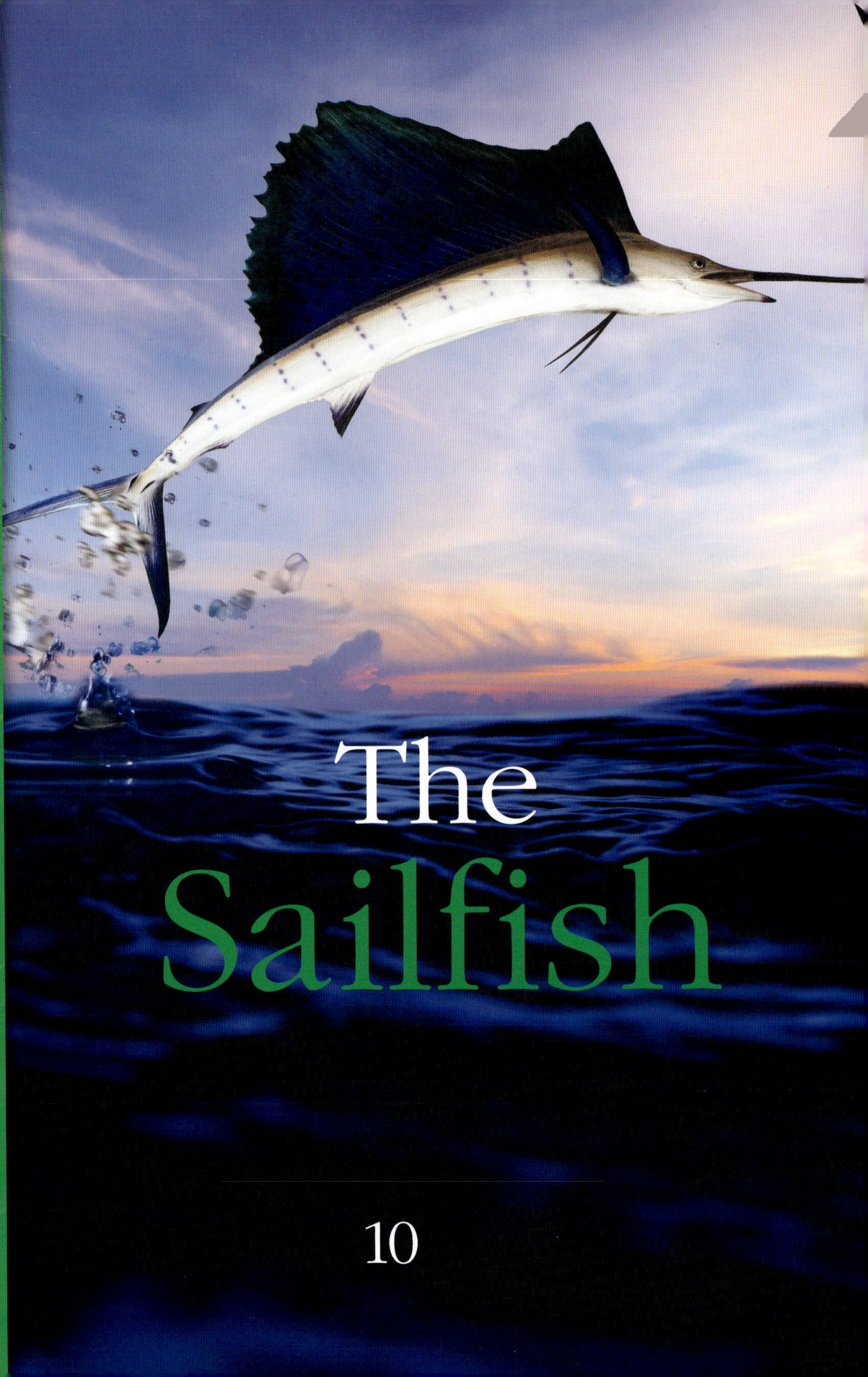

3

The Sailfish

Out of all the fish in the sea, the sailfish is the fastest. It can swim up to 68 miles (109 km) per hour. One thing that makes the sailfish so fast is its ability to jump. It leaps out of the water while swimming.

Sailfish are also one of the biggest saltwater fish. A large sailfish can be up to 11 feet (3.4 meters) long. It weighs up to 220 pounds (100 kilograms). Its long bill slashes fish swimming in a school. This makes it easier for the sailfish to catch and eat the fish.

Did You Know?

Sailfish lift their back fin out of the water when attacking fish.

4

The Peregrine Falcon

The peregrine falcon can fly as fast as 69 miles (111 km) per hour. Falcons eat other birds. While hunting, falcons dive bomb the birds at speeds reaching 200 miles (322 km) per hour. That's as fast as a racecar!

To reach top speeds, the peregrine falcon flies very high in the sky. Then it dives. Male peregrines also show off for females by flying high.

Did You Know?
Peregrine falcons can live up to 17 years in the wild.

5

The Dragonfly

Zoom! Something flies by in a blur. It's a dragonfly! Dragonflies are the fastest flying insects. They reach up to 34 miles (55 km) per hour.

Dragonflies are fast and strong fliers. They have two sets of wings. The wings move fast. Humans cannot see them. Dragonflies can fly forward and backward. They can change direction in midair. They can **hover** in the air for up to one minute.

Dragonflies use their speed to hunt. They are one of the top predators of the insect world. They will eat almost any bug they can catch.

Did You Know?
One dragonfly can eat hundreds of mosquitoes in a day.

6

The Mite

Mites are tiny arachnids. They are no bigger than a sesame seed. Some are so small you need a microscope to see them.

Mites are fast. Compared to their size, they cover more ground than any other animal. If a mite were human-sized, it could run 1,228 miles (1,976 km) per hour! Talk about speed!

Mites can stop and turn quickly. They can even run on sizzling hot ground. Their speed and size make them hard to study. Scientists are still learning what mites eat.

Think About It
Do you think the mite should be named the fastest animal?

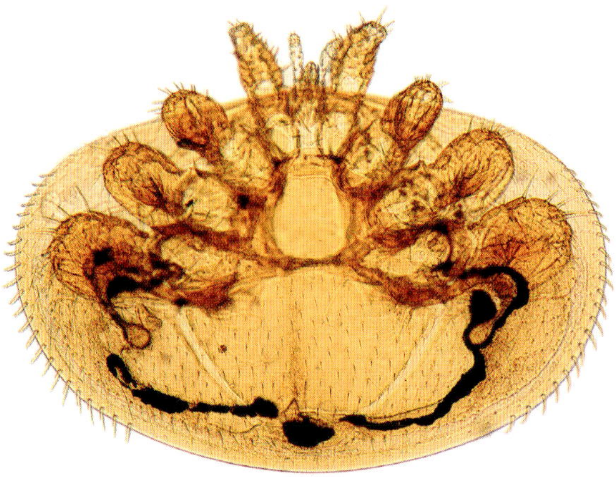

MORE TO EXPLORE

WHERE IN THE WORLD?

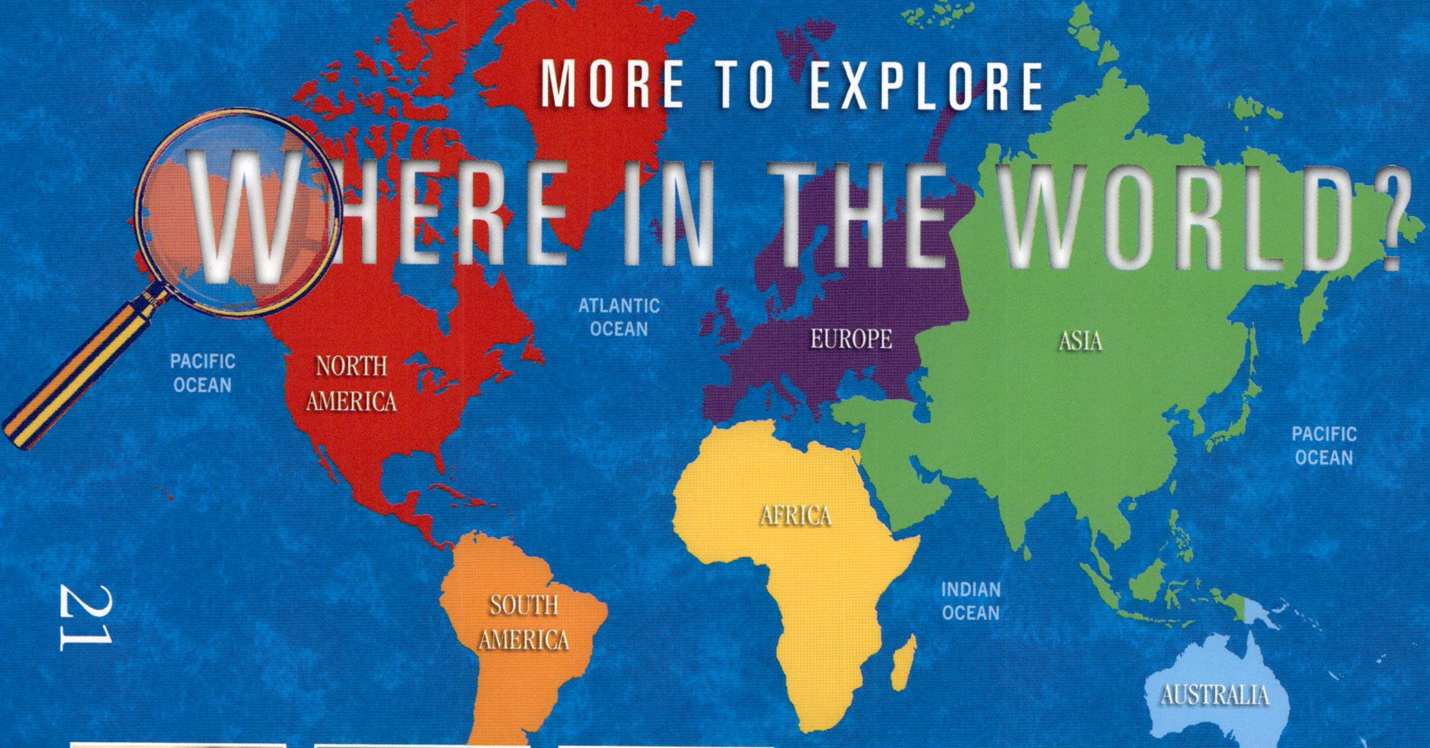

CHEETAHS
Africa

DRAGONFLIES
All over the world except Antarctica

MITES
All over the world except Antarctica

PEREGRINE FALCONS
All over the world except Antarctica

PRONGHORNS
North America

SAILFISH
Warm and temperate parts of all oceans

MORE TO EXPLORE

FANTASTIC FACTS

Cheetahs have about 2,000 spots.

Some people like to cook and eat dragonflies for a snack.

Pronghorns have great eyes. They can spot predators from very far away.

Sometimes, mites live on an animal's skin.

Peregrine falcons mostly eat other birds.

Sailfish are smart. They often get together to hunt fish.

MORE TO EXPLORE

COOL COMPARISONS

What are the top speeds of each animal?

Dragonflies
34 miles (55 km) per hour

Mites
0.7 miles (1.1 km) per hour

Peregrine Falcons
200 miles (322 km) per hour

Pronghorns
60 miles (97 km) per hour

Cheetahs
75 miles (121 km) per hour

Sailfish
68 miles (109 km) per hour

MORE TO EXPLORE

RESOURCES

Glossary

arachnid (uh-RAK-nid) A class of animals including spiders, scorpions, mites, and ticks.

hover (HUHV-er) Floating in the air without moving in any direction.

predator (PRED-uh-tuhr) An animal that eats other animals.

school (SKOOL) A group of fish.

spine (SPINE) A backbone.

Read More

Austen, Lily. *Fastest Animals*. Minneapolis: Jump!, Inc., 2025.

Riggs, Kate. *Falcons*. Mankato, MN: The Creative Company, 2023.

Index

TOP RANK is published by Black Rabbit Books, P.O. Box 227, Mankato, MN, 56002. • COPYRIGHT © 2025 Black Rabbit Books. All rights reserved. No part of this book may be reproduced in any form without written permission from the publisher. • Top Rank is an imprint of Black Rabbit Books. • Edited by Alissa Thielges • Designed by Danny Nanos • Photographs © Alamy Stock Photo/Rodrigo Friscione, 12; Dreamstime/Maksim Gavrilenko, cover; Getty Images/FRANKHILDEBRAND, 13, 21, 23, olikim, 20; iStock/rpbirdman, 6–7; Shutterstock/Alexander Raths, 11, archivector, 23, Darkdiamond67, 17, Dennis W Donohue, 9, 21, Efimova Anna, 4, 21, Elana Erasmus, 5, evgo1977, 23, Harry Collins Photography, 2, 14–15, Irina Simkina, 23, Jim H Walling, 19, 21, Klimek Pavol, 18, Kozyreva Elena, 23, Moment of Perception, 8, Stock High angle view, 11, stockphoto mania, 10, 21, Svetlana Foote, 5, Viktorya170377, 23, vnlit, cover, yanikap, 16, 21 • Printed in China

Library of Congress Cataloging-in-Publication Data: Names: Greve, Meg, author. | Title: Super-fast animals / by Meg Greve. | Description: Mankato, MN : Top Rank is an imprint of Black Rabbit Books, [2025] | Series: Super-incredible animals | Includes bibliographical references and index. | Audience: Ages 8–11 | Audience: Grades 2–3 | Identifiers: LCCN 2024017496 | ISBN 9781645824008 (paperback) | ISBN 9781645823780 (library binding) | ISBN 9781645824220 (ebook) | Subjects: LCSH: Animal locomotion—Juvenile literature. | Speed—Juvenile literature. | Classification: LCC QP301 .G684 2025 | DDC 591.47/9—dc23/eng/20240516 | LC record available at https://lccn.loc.gov/2024017496